ENERGY SECTOR STANDARD
OF THE PEOPLE'S REPUBLIC OF CHINA

中华人民共和国能源行业标准

Code for Hydrological Forecasting
of Hydropower Projects

水电工程水文预报规范

NB/T 10085-2018

Chief Development Department: China Renewable Energy Engineering Institute

Approval Department: National Energy Administration of the People's Republic of China

Implementation Date: March 1, 2019

China Water & Power Press

中国水利水电出版社

Beijing 2024

All rights reserved. No part of this publication may be reproduced, stored in a retrieval system, or transmitted in any form or by any means—electronic, mechanical, photocopying, recording or otherwise, without prior written permission of the publisher.

图书在版编目（CIP）数据

水电工程水文预报规范：NB/T 10085-2018 = Code for Hydrological Forecasting of Hydropower Projects(NB/T 10085-2018)：英文 / 国家能源局发布.
北京：中国水利水电出版社，2024. 10. -- ISBN 978-7-5226-2768-7
Ⅰ. TV-65
中国国家版本馆CIP数据核字第2024PJ8966号

ENERGY SECTOR STANDARD
OF THE PEOPLE'S REPUBLIC OF CHINA
中华人民共和国能源行业标准

Code for Hydrological Forecasting
of Hydropower Projects
水电工程水文预报规范
NB/T 10085-2018
（英文版）

Issued by National Energy Administration of the People's Republic of China
国家能源局　发布
Translation organized by China Renewable Energy Engineering Institute
水电水利规划设计总院　组织翻译
Published by China Water & Power Press
中国水利水电出版社　出版发行
　　Tel: (+ 86 10) 68545888　68545874
　　sales@mwr.gov.cn
　　Account name: China Water & Power Press
　　Address: No.1, Yuyuantan Nanlu, Haidian District, Beijing 100038, China
　　http://www.waterpub.com.cn
中国水利水电出版社微机排版中心　排版
北京中献拓方科技发展有限公司　印刷
184mm×260mm　16开本　2.5印张　79千字
2024年10月第1版　2024年10月第1次印刷
Price(定价)：￥360.00

Introduction

This English version is one of China's energy sector standard series in English. Its translation was organized by China Renewable Energy Engineering Institute authorized by National Energy Administration of the People's Republic of China in compliance with relevant procedures and stipulations. This English version was issued by National Energy Administration of the People's Republic of China in Announcement [2022] No. 4 dated May 13, 2022.

This version was translated from the Chinese Standard NB/T 10085-2018, *Code for Hydrological Forecasting of Hydropower Projects*, published by China Water & Power Press. The copyright is reserved by National Energy Administration of the People's Republic of China. In the event of any discrepancy in the implementation, the Chinese version shall prevail.

Many thanks go to the staff from relevant standard development organizations and those who have provided generous assistance in the translation and review process.

For further improvement of the English version, any comments and suggestions are welcome and should be addressed to:

China Renewable Energy Engineering Institute
No. 2 Beixiaojie, Liupukang, Xicheng District, Beijing 100120, China
Website: www.creei.cn

Translating organization:

POWERCHINA Zhongnan Engineering Corporation Limited

China Renewable Energy Engineering Institute

Translating staff:

LI Qian   CHEN Lei   YANG Hong   GAO Jie

Review panel members:

GUO Jie	POWERCHINA Beijing Engineering Corporation Limited
QIE Chunsheng	Senior English Translator
LIU Xiaofen	POWERCHINA Zhongnan Engineering Corporation Limited
YAN Wenjun	Army Academy of Armored Forces, PLA
ZHANG Ming	Tsinghua University

YAN Cunku	POWERCHINA Northwest Engineering Corporation Limited
YOU Yang	China Society for Hydropower Engineering
WANG Tao	POWERCHINA Chengdu Engineering Corporation Limited
WAN Wengong	China Renewable Energy Engineering Institute
YAN Zhonglin	POWERCHINA Zhongnan Engineering Corporation Limited
LI Shisheng	China Renewable Energy Engineering Institute

National Energy Administration of the People's Republic of China

翻译出版说明

本译本为国家能源局委托水电水利规划设计总院按照有关程序和规定，统一组织翻译的能源行业标准英文版系列译本之一。2022年5月13日，国家能源局以2022年第4号公告予以公布。

本译本是根据中国水利水电出版社出版的《水电工程水文预报规范》NB/T 10085—2018翻译的，著作权归国家能源局所有。在使用过程中，如出现异议，以中文版为准。

本译本在翻译和审核过程中，本标准编制单位及编制组有关成员给予了积极协助。

为不断提高本译本的质量，欢迎使用者提出意见和建议，并反馈给水电水利规划设计总院。

地址：北京市西城区六铺炕北小街2号
邮编：100120
网址：www.creei.cn

本译本翻译单位：中国电建集团中南勘测设计研究院有限公司
　　　　　　　　水电水利规划设计总院

本译本翻译人员：李　倩　陈　蕾　杨　虹　高　洁

本译本审核人员：

郭　洁　中国电建集团北京勘测设计研究院有限公司

郄春生　英语高级翻译

刘小芬　中国电建集团中南勘测设计研究院有限公司

闫文军　中国人民解放军陆军装甲兵学院

张　明　清华大学

严存库　中国电建集团西北勘测设计研究院有限公司

由　洋　中国水力发电工程学会

王　涛　中国电建集团成都勘测设计研究院有限公司

万文功　水电水利规划设计总院

晏忠林　中国电建集团中南勘测设计研究院有限公司

李仕胜　水电水利规划设计总院

国家能源局

Announcement of National Energy Administration of the People's Republic of China
[2018] No. 12

According to the requirements of Document GNJKJ [2009] No. 52, "Notice on Releasing the Energy Sector Standardization Administration Regulations (*tentative*) and detailed implementation rules issued by National Energy Administration of the People's Republic of China", 204 sector standards such as *Coal Mine Air-Cooling Adjustable-Speed Magnetic Coupling*, including 54 energy standards (NB), 8 petrochemical standards (NB/SH), and 142 petroleum standards (SY), are issued by National Energy Administration of the People's Republic of China after due review and approval.

Attachment: Directory of Sector Standards

National Energy Administration of the People's Republic of China

October 29, 2018

Attachment:

Directory of Sector Standards

Serial number	Standard No.	Title	Replaced standard No.	Adopted international standard No.	Approval date	Implementation date
...						
40	NB/T 10085-2018	Code for Hydrological Forecasting of Hydropower Projects			2018-10-29	2019-03-01
...						

Foreword

According to the requirements of Document GNKJ [2015] No. 12 issued by the National Energy Administration of the People's Republic of China, "Notice on Releasing the Development and Revision Plan of the Second Batch of Energy Sector Standards in 2014", and after extensive investigation and research, summarization of practical experience, and wide solicitation of opinions, the drafting group has prepared this code.

The main technical contents of this code include: basic data, flood forecasting, runoff forecasting, ice forecasting and tide level forecasting, hydrological forecasting during river closure and reservoir impoundment, rating criteria for hydrological forecasting schemes, and operational forecasting.

National Energy Administration of the People's Republic of China is in charge of the administration of this code. China Renewable Energy Engineering Institute has proposed this code and is responsible for its routine management. The Energy Sector Standardization Technical Committee on Hydropower Planning, Resettlement and Environmental Protection is responsible for the explanation of specific technical contents. Comments and suggestions in the implementation of this code should be addressed to:

China Renewable Energy Engineering Institute
No. 2 Beixiaojie, Liupukang, Xicheng District, Beijing 100120, China

Chief development organization:

China Renewable Energy Engineering Institute

POWERCHINA Zhongnan Engineering Corporation Limited

Participating development organization:

POWERCHINA Chengdu Engineering Corporation Limited

Yalong River Hydropower Development Company, Ltd.

China Yangtze Power Co., Ltd.

Chief drafting staff:

YAN Zhonglin	YANG Baiyin	ZHANG Yang	GAO Jie
FEI Rujun	LIU Jingyi	WANG Tao	HUANG Peijin
YU Hao	CAO Yuanyuan	LI Ming	HE Zhaohui
LIU Zhiwu	LIANG Jun		

Review panel members:

WAN Wengong	ZHANG Min	ZHOU Xinchun	LIU Guangbao
CAO Changxiang	YE Xugang	XIA Jianrong	ZHAO Jimin
YANG Zhongmin	LI Xiaowei	YU Fuliang	SHI Peng
WANG Yuhua	DONG Xianyong	CAI Pin	PIAO Ling
LI Shisheng			

Contents

1	**General Provisions**	1
2	**Terms**	2
3	**Basic Data**	3
4	**Flood Forecasting**	4
4.1	General Requirements	4
4.2	Contents and Methods of Flood Forecasting Schemes	5
4.3	Rating and Verification of Flood Forecasting Schemes	7
4.4	Applicability of Flood Forecasting Schemes	8
5	**Runoff Forecasting**	9
5.1	General Requirements	9
5.2	Contents and Methods of Runoff Forecasting Schemes	9
5.3	Rating and Verification of Runoff Forecasting Schemes	10
5.4	Applicability of Runoff Forecasting Schemes	10
6	**Ice Forecasting and Tide Level Forecasting**	12
6.1	General Requirements	12
6.2	Ice Forecasting	12
6.3	Tide Level Forecasting	12
7	**Hydrological Forecasting During River Closure and Reservoir Impoundment**	13
7.1	Hydrological Forecasting During River Closure	13
7.2	Hydrological Forecasting During Reservoir Impoundment	13
8	**Rating Criteria for Hydrological Forecasting Schemes**	14
8.1	Forecast Error, Permissible Forecast Error and Forecast Qualified Rate	14
8.2	Rating Criteria for Flood Forecasting Schemes	14
8.3	Rating Criteria for Runoff Forecasting Schemes	15
8.4	Rating Criteria for Ice Forecasting Schemes	16
8.5	Rating Criteria for Tide Level Forecasting Schemes	17
9	**Operational Forecasting**	19
9.1	Hydrological Station Network and Water Regime Data	19
9.2	Requirements for Organization and Staffing	19
9.3	Software Systems and Forecasting Operations	20
9.4	Information Reporting and Dissemination	21
9.5	Forecast Accuracy Evaluation	21
9.6	Forecast Analysis and Summary	22
Appendix A	**Contents of a Flood Forecasting Scheme**	23

Appendix B Contents of a Runoff Forecasting Scheme ········ 24
Explanation of Wording in This Code ···························· 25
List of Quoted Standards ·· 26

1 General Provisions

1.0.1 This code is formulated with a view to specifying the work content, methods and technical requirements for hydrological forecasting of hydropower projects.

1.0.2 This code is applicable to the hydrological forecasting of hydropower projects.

1.0.3 Hydrological forecasting during the construction period of a hydropower project shall meet the requirements for flood protection during construction and reservoir impoundment; hydrological forecasting during the operation period of a hydropower project shall meet the requirements for flood protection, reservoir operation and electric power market.

1.0.4 The hydrological forecasting for hydropower projects shall mainly include flood forecasting, runoff forecasting, ice forecasting, and tide level forecasting.

1.0.5 In addition to this code, the hydrological forecasting of hydropower projects shall comply with other current relevant standards of China.

2 Terms

2.0.1 hydrological model

physical model or mathematical and logical model developed to simulate hydrological phenomena

2.0.2 forecasting element

physical quantity to forecast the water regime at a specific location or region at a given time, such as flow rate, water level, flow velocity, and ice thickness

2.0.3 forecast lead time

length of time between the data time for forecasting and the occurrence time of the forecasting element

2.0.4 permissible forecast error

permissible range of forecast error determined comprehensively based on the uses of forecast result and the technological level of forecasting

2.0.5 deterministic coefficient

factor to indicate the extent to which the forecast flood process agrees with the measured flood process

2.0.6 data time for forecasting

final measurement time of hydrological data serving as the basis for hydrological forecast

2.0.7 qualified forecast

forecast whose error falls within the permissible range specified for the forecasting element

2.0.8 forecast qualified rate

percentage of qualified forecasts in the total forecasts for a forecasting element

3 Basic Data

3.0.1 The basic data used for formulating the forecasting scheme shall mainly include:

   1   Hydrometeorological data, including water level, flow, precipitation, evaporation, ice regime and tidal level.

   2   Data on underlying surface, including the landform, vegetation, groundwater level, and thickness of aeration zone in the river basin.

   3   Related project information, including the project construction progress and the upstream and downstream cascade development.

3.0.2 The hydrometeorological data used for formulating the forecasting scheme shall be reliable, representative and consistent.

3.0.3 The hydrological basis stations used for formulating the forecasting scheme shall be the same ones for operational forecasting, and a database shall be established according to the requirements for the structure and identifiers of hydrological database for hydropower projects.

3.0.4 Data series used in formulating the flood forecasting scheme shall not be less than 10 years, which shall cover the typical years of large, medium and small floods. The number of typical flood events shall be no less than 50 for humid regions and no less than 25 for arid regions. The data of floods higher than the median of sample flood peaks shall all be adopted. In the case of data shortage, all the flood data shall be used.

3.0.5 Data series shall be no less than 10 years for formulating the short-term runoff forecasting scheme and no less than 30 years for formulating the medium- and long-term runoff forecasting scheme. In the case of data shortage, all the runoff data shall be used.

3.0.6 The number of typical samples for formulating the ice forecasting scheme shall be no less than 30. In the case of sample shortage, all the ice regime data shall be used.

3.0.7 In formulating the tide forecasting scheme, tide level data of no less than 10 tropical (temperature zone) cyclone events shall be used for the surge forecasting scheme, and the hourly tide level data of no less than 1 year, including the high tide level, low tide level and their occurrence times, shall be used for the normal tide level forecasting scheme. In the case of data shortage, all the tide level data shall be used.

4 Flood Forecasting

4.1 General Requirements

4.1.1 The flood forecasting for a control cross section or key location of a river shall mainly cover the following flood elements:

1. The peak flood discharge, peak flood level, flood peak occurrence time and flood process at the control cross section or key location of the river.

2. The peak flood discharge, flood peak occurrence time and flood process at the dam site during construction; the processes, values and occurrence time of peak water level and flow at the inlet and outlet of a diversion tunnel, the upstream and downstream cofferdams and other key locations in construction area.

3. The discharge and occurrence time of peak inflow flood, flow hydrograph and flood volume, and the reservoir water level hydrograph and reservoir outflow hydrograph during operation.

4.1.2 In flood forecasting during project construction period, a forecast cross section, its forecasting elements, forecast period and forecast lead time shall be determined according to the flood protection requirements and the hydrological characteristics of the river basin.

4.1.3 In flood forecasting during project operation period, a forecast cross section, its forecasting elements, forecast period and forecast lead time shall be determined according to the flood protection requirements and reservoir operation and the hydrological characteristics of the river basin.

4.1.4 A flood forecasting scheme shall be developed according to forecasting elements and forecast cross sections. The longest forecast lead time shall be adopted provided that the forecast accuracy requirements are satisfied.

4.1.5 The forecasting scheme shall be subjected to rating and verification to determine its class. In either of the following cases, the forecasting scheme shall be degraded.

1. Data series is insufficient.

2. The verification class is lower than the rating class, the influence factor is not identified and no additional data can be used for re-verification.

4.1.6 For operational forecasting using rainfall forecasting results, the forecasting basis and reliability shall be stated when the forecast is disseminated.

4.1.7 For the project with its construction area prone to flash flood, a flash flood forewarning and forecasting scheme shall be developed.

4.2 Contents and Methods of Flood Forecasting Schemes

4.2.1 The formulation of a flood forecasting scheme shall:

1. Analyze the hydrometeorological characteristics of the river basin and the upstream and downstream project characteristics, and get an understanding of the hydrological and meteorological station network in the river basin.

2. Analyze the requirements of project construction or operation for flood forecasting.

3. Analyze the impact of cascade reservoir operation on flood forecasting scheme.

4. Collect and compile the flow and water level series and rating curve of related cross section, the rainfall and evaporation series of the forecast river basin, reservoir storage curve and outflow curve, etc.

5. Analyze the average flood travel time in the river reaches between an upstream hydrological station and the forecast cross section to determine the forecast lead time of each cross section.

6. Select the hydrological forecasting method.

7. Analyze and determine the forecasting scheme, which shall cover inputs and outputs, model combination, calculation method, etc.

8. Debug the model and calibrate the parameters.

9. Rate and verify the forecasting scheme.

4.2.2 Flood forecasting should use the flood forecasting models or methods that are well proven, advanced, reliable, and widely applicable. Xin'anjiang Model, Shanbei Model or SWAT Model may be selected as the forecasting model.

4.2.3 The runoff yield and flow concentration calculation for a forecasting scheme shall follow the water budget principle and meet the following requirements:

1. In runoff yield calculation of a river basin, a proper runoff yield model should be established according to the climatic and geographical conditions and underlying surface of the river basin; the rainfall-

runoff relationship should be established according to the rainfall volume, evaporation and flow at the forecast cross section, to calculate the runoff yield volume of the forecast river basin. In the absence of measured data, regional generalized method or hydrological analogy may be used in runoff yield calculation.

2 In flow concentration calculation of a river basin, the flood hydrograph of the forecast cross section shall be calculated with hydrological or hydrodynamic flow concentration model according to the runoff yield calculation result and the measured flow of the cross section. In the absence of measured data, the regional generalized method or the empirical formula may be used in flow concentration calculation.

3 For a river basin with comparatively large area, uneven rainfall distribution or greater difference in runoff yield and flow concentration conditions, the river basin should be divided into sub-catchments for runoff yield and flow concentration calculation.

4 An integrated model or a combination of multiple models may be used for the runoff yield and flow concentration calculation.

4.2.4 The formulation of flood forecasting schemes shall meet the following requirements:

1 A number of methods shall be selected to formulate a flood forecasting scheme according to the characteristics of runoff yield and flow concentration in the river basin, the pattern of flood routing in river course, and the water and rainfall regime monitoring and flood forecasting conditions.

2 The selected flood forecasting method shall be able to realize the forecast of expected flood elements and achieve the specified forecast accuracy.

3 The impact of the operation of upstream cascade reservoirs shall be considered.

4 Parameter calibration for the flood forecasting model shall analyze the sensitivity, rationality and reliability of the parameters as well as the stability of the model by optimization calculation and rationality analysis.

5 Flood routing shall be carried out after water is retained by a cofferdam or dam to forecast the flood elements such as the water level in front of the cofferdam or dam and the outflow.

6　The reservoir flood forecasting scheme during operation period shall cover the flood hydrograph forecasting and the forecasting of such flood elements as the water level in front of the dam and outflow according to the reservoir operation scheme.

4.2.5　Each year after flood season, the flood forecasting scheme shall be assessed.

4.2.6　The flood forecasting scheme shall be revised, supplemented or updated in one of the following cases:

1　The measured hydrological data has exceeded the extreme values of the data used in formulation of the flood forecasting scheme.

2　The water regime of the river basin, river reach or cross section has changed due to the change in natural conditions or the influence of human activities.

3　Changes have occurred to the hydrological basis station used for flood forecasting.

4　Forecast accuracy may be improved or forecast lead time may be extended using new methods or technologies.

4.2.7　The flash flood disaster forewarning and forecasting scheme for project area should be developed using the flash flood disaster analysis and assessment technique. Forewarning indicators should be proposed, which should be revised and adjusted timely according to the actual applications.

4.2.8　The contents of a flood forecasting scheme shall be in accordance with Appendix A of this code.

4.3　Rating and Verification of Flood Forecasting Schemes

4.3.1　Flood forecasting schemes shall be classified into Class A, Class B and Class C.

4.3.2　The rating items shall include peak flood discharge, peak flood level, flood peak occurrence time, flood volume or runoff depth, flood hydrograph, etc.

4.3.3　The rating indicators shall be categorized by forecasting element. Qualified rate shall be used for peak flood discharge, peak flood level, flood peak occurrence time, and flood volume or runoff depth. Deterministic coefficient shall be used for flood hydrograph.

4.3.4　The rating of a flood forecasting scheme shall use all the data applied in modeling.

4.3.5 The flood forecasting scheme shall be verified using no less than 2-year data that have not been applied in modeling. The indicators for verifying the flood forecasting scheme shall be the same as those for rating the scheme.

4.3.6 When the verified class of a flood forecasting scheme is lower than the rated class, the cause shall be analyzed, and additional data should be used for re-modeling, re-rating and re-verification. If the rated class of the flood forecasting scheme is inconsistent with the verified class, the lower class shall apply.

4.4 Applicability of Flood Forecasting Schemes

4.4.1 The flood forecasting scheme of Class A or Class B shall be used for issuing a formal forecast.

4.4.2 The flood forecasting scheme of Class C shall be used for issuing a reference forecast.

4.4.3 The flood forecasting scheme below Class C shall not be used for issuing a formal forecast or a reference forecast.

5 Runoff Forecasting

5.1 General Requirements

5.1.1 Runoff forecasting may be classified by forecast period into the short-term runoff forecasting and the medium- and long-term runoff forecasting on a ten-day, monthly, quarterly or yearly basis.

5.1.2 In runoff forecasting, the runoff volume and runoff magnitude at the forecast cross section in a coming time period shall be forecast. The forecasting results may cover runoff advent, runoff series, and qualitative runoff magnitude.

5.1.3 In the formulation of a runoff forecasting scheme, the runoff forecast period and forecasting results shall be determined according to the requirements of project construction and operation.

5.1.4 The runoff forecasting scheme shall be subjected to rating and verification to determine its class.

5.1.5 The runoff forecasting may use the rainfall data of meteorological forecast. For the operational forecasting using such rainfall data, the forecasting basis and reliability shall be stated when the forecast is issued.

5.2 Contents and Methods of Runoff Forecasting Schemes

5.2.1 The formulation of a runoff forecasting scheme shall:

1. Analyze the hydrometeorological characteristics of the river basin and the characteristics of the upstream and downstream projects concerned, and understand the basic conditions of the hydrometeorological station network in the river basin concerned.

2. Define the requirements of project construction or operation for runoff forecasting.

3. Analyze and determine the runoff forecasting period.

4. Select the runoff forecasting method. The selected method shall ensure that the data used as the basis of operational forecasting meet the requirements.

5. Collect and process other data required for formulating the forecasting scheme.

6. Analyze the effect of the upstream cascade reservoirs on the runoff forecasting.

7   Analyze and determine the forecasting scheme, including the input and output, the model combinations, and the calculation method.

8   Debug the models and calibrate parameters.

9   Rate and verify the runoff forecasting scheme.

5.2.2   Runoff forecasting should use the runoff forecasting models or methods that are well proven, advanced, reliable, and widely applicable. The short-term runoff forecasting may use the flow hydrograph of flood forecasting. The ensemble forecasting method should be used for runoff forecasting to determine the forecasting results in a comprehensive way.

5.2.3   The contents of a runoff forecasting scheme shall be in accordance with Appendix B of this code.

5.3   Rating and Verification of Runoff Forecasting Schemes

5.3.1   Runoff forecasting schemes shall be classified into Class A, Class B and Class C.

5.3.2   Runoff forecasting schemes shall be rated in terms of short-term runoff forecasting, and medium- and long-term runoff forecasting.

5.3.3   The forecast qualified rate shall be used as the rating indicator for runoff forecasting schemes.

5.3.4   Runoff forecasting scheme rating shall use all the data applied in modeling.

5.3.5   The runoff forecasting scheme shall be verified using no less than 2-year data that have not been applied in modeling for the forecasting scheme. The indicators for verifying the runoff forecasting scheme shall be the same as those for rating the scheme.

5.3.6   When the verified class of a runoff forecasting scheme is lower than its rated class, the cause shall be analyzed. Additional data should be used for re-modeling, re-rating and re-verification. If the rated class of the runoff forecasting scheme is inconsistent with the verified class, the lower class shall apply.

5.4   Applicability of Runoff Forecasting Schemes

5.4.1   For short-term runoff forecasting schemes, Class A and Class B schemes shall be used for formal forecasting; Class C schemes shall be used for reference forecasting; schemes below Class C shall not be used for forecasting.

5.4.2   For medium- or long-term runoff forecasting schemes, Class A and

Class B schemes shall be used for formal forecasting; Class C schemes shall be used for reference forecasting.

5.4.3 Medium- and long-term qualitative runoff forecasting schemes shall be used for reference forecasting.

6 Ice Forecasting and Tide Level Forecasting

6.1 General Requirements

6.1.1 For hydropower projects subject to freezing, ice forecasting should be carried out; for tidal power stations and hydropower projects subject to tidal effects, tide level forecasting should be carried out.

6.1.2 For ice or tide level forecasting schemes, Class A and Class B schemes shall be used for formal forecasting; Class C schemes shall be used for reference forecasting.

6.1.3 For ice or tide level forecasting schemes, the forecast qualified rate shall be used as the rating indicator.

6.2 Ice Forecasting

6.2.1 Ice forecasting shall include freezing forecasting and thawing forecasting. The freezing forecasting items shall include the channel storage volume, flowing ice date, freezing date, ice thickness, maximum ice volume of the river reach, and flowing ice volume at the cross section. For unstable frozen river reaches, the freezing regime shall be included as well. The thawing forecasting items shall include the thawing date and thawing regime.

6.2.2 In formulating an ice forecasting scheme, the empirical method, the statistical method, and the ice regime mathematical model based on hydraulic and thermal conditions of the river course may be adopted. The meteorological and hydrological forecasting factors used in the ice regime mathematical model shall conform to the physical origin of ice regime, and rationality analysis shall be carried out on the applied model and the forecasting results.

6.3 Tide Level Forecasting

6.3.1 Tide level forecasting shall cover the coastal area subject to astronomic tide and storm tide and the estuary and tidal reach under the action of river flow, the rising of astronomic tide and the surge of storm tide. The tide level forecasting items shall include normal tide level, surge, and maximum tide level and its occurrence time.

6.3.2 The empirical and/or numerical methods may be selected for tide level forecasting depending on the actual conditions. The forecast lead time shall not be less than 6 h.

7 Hydrological Forecasting During River Closure and Reservoir Impoundment

7.1 Hydrological Forecasting During River Closure

7.1.1 The hydrological forecasting during river closure shall include flood forecasting and runoff forecasting.

7.1.2 The scheme for hydrological forecasting during river closure shall be formulated according to flood and runoff forecasting methods and the river closure progress, and shall cover the hydrological elements such as the water levels at the inlet and outlet of the diversion tunnel and the upstream and downstream of the cofferdam and the flow velocity, hydraulic drop or backwater height at the closure gap.

7.1.3 The scheme for hydrological forecasting during river closure and forecast results shall be timely corrected or modified according to the actual river closure progress, the real-time monitored upstream water level, the width and flow velocity at the closure gap, and the change in water diversion ratio.

7.2 Hydrological Forecasting During Reservoir Impoundment

7.2.1 The hydrological forecasting during reservoir impoundment shall include flood forecasting and runoff forecasting.

7.2.2 Reservoir flood routing shall be carried out according to the reservoir storage curve and the forecast results of the reservoir inflow, to predict the hydrograph of the water level in front of the dam.

8 Rating Criteria for Hydrological Forecasting Schemes

8.1 Forecast Error, Permissible Forecast Error and Forecast Qualified Rate

8.1.1 The forecast error shall be expressed in terms of absolute forecast error or relative forecast error.

8.1.2 The permissible forecast error shall be defined for each forecasting element.

8.1.3 A qualified forecast shall meet the following requirements:

1. For the forecast results of numerical type, the forecast error is not greater than the permissible forecast error.

2. For the forecast results of discrete type, the forecast results agree with the characteristics of actual events.

8.1.4 The forecast qualified rate shall be calculated by the following formula:

$$QR = \frac{n}{m} \times 100 \ \% \tag{8.1.4}$$

where

- QR is the forecast qualified rate, rounded to one decimal place;
- n is the number of qualified forecasts;
- m is the total number of forecasts.

8.2 Rating Criteria for Flood Forecasting Schemes

8.2.1 In rating a flood forecasting scheme, the forecast effect for peak flood discharge, peak flood level, flood peak occurrence time, flood volume or runoff depth, and flood hydrograph of each flood event shall be assessed.

8.2.2 The permissible forecast errors of flood forecasts shall be taken according to the following requirements:

1. The permissible forecast error of the peak flood discharge forecast using rainfall runoff takes 20 % of the measured peak flood discharge.

2. The permissible forecast errors of the peak flood discharge and peak flood level forecast using the discharge and water level of the river course takes 20 % of the variation in the measured discharge and water level within the forecast lead time. When the permissible forecast error of discharge is less than 5 % of the measured discharge, it takes 5 % of

the measured discharge; when the permissible forecast error of water level is less than the water level amplitude corresponding to 5 % of the measured peak flood discharge, it takes the water level amplitude corresponding to 5 % of the measured peak flood discharge and is no less than 0.10 m.

3   The flood peak occurrence time takes 30 % of the time period from the data time for forecasting to the occurrence time of the flood peak. When the time is less than 3 h or a forecast period, it takes 3 h or the forecast period.

4   The permissible forecast error of runoff depth using the rainfall-runoff forecasting model takes 20 % of the measured runoff depth and is controlled with an upper limit of 20 mm and a lower limit of 3 mm.

5   The permissible forecast error of flow velocity takes 20 % of the measured flow velocity.

8.2.3   In rating a flood forecasting scheme, the forecast qualified rate is employed as the rating indicator for peak flood discharge, peak flood level, flood peak occurrence time and runoff depth; the deterministic coefficient is employed as the rating indicator for the flood hydrograph forecasting. The rating criteria for flood forecasting schemes are given in Table 8.2.3.

Table 8.2.3   Rating criteria for flood forecasting schemes

Class	A	B	C
Forecast qualified rate QR	$QR \geq 85.0\%$	$85.0\% > QR \geq 70.0\%$	$70.0\% > QR \geq 60.0\%$
Deterministic coefficient DC	$DC \geq 0.90$	$0.90 > DC \geq 0.70$	$0.70 > DC \geq 0.50$

8.2.4   The class of a flood forecasting scheme shall be determined based on the comprehensive consideration of various forecasting elements. When the class of the dominant forecasting element is lower than that of the scheme, the class of the dominant forecasting element shall prevail.

8.3   Rating Criteria for Runoff Forecasting Schemes

8.3.1   Runoff forecasting scheme rating shall assess the runoff volume forecasting effect.

8.3.2   The permissible forecast error of runoff volume shall not exceed 20 % of the measured runoff volume.

8.3.3   Runoff forecasting scheme rating shall use the forecast qualified rate as the rating indicator. The class of a runoff forecasting scheme in each forecast period shall be determined according to Table 8.3.3, and the rating result of the advent period shall be taken as the final rating result of the forecasting scheme. The rating class of runoff series forecasting in other periods shall be used as the reference for forecasting scheme rating.

Table 8.3.3 Rating criteria for runoff forecasting scheme

Class	A	B	C
Forecast qualified rate QR	$QR \geq 85.0\%$	$85.0\% > QR \geq 70.0\%$	$70.0\% > QR \geq 60.0\%$

8.3.4 A medium- and long-term qualitative runoff forecast shall be rated by the measured runoff volume which shall be classified as high or low flow according to Table 8.3.4. When the forecast runoff conforms to the indicator for classification of flow level by measured runoff, it is considered "qualified"; otherwise "unqualified". The medium- and long-term qualitative runoff forecasting scheme may be subjected to statistics on the qualified rate, but should not be rated.

Table 8.3.4 Classification of flow level by measured runoff

Flow level	Low	Comparatively low	Normal	Comparatively high	High
Runoff anomaly	$< -20\%$	$-20\% \leq$ anomaly $< -10\%$	$-10\% \leq$ anomaly $\leq 10\%$	$10\% <$ anomaly $\leq 20\%$	$> 20\%$

8.4 Rating Criteria for Ice Forecasting Schemes

8.4.1 Ice forecasting scheme rating shall include the forecasting effect for the forecasting elements and their occurrence time.

8.4.2 The permissible forecast error of ice forecasting elements shall take 25 % of the variation amplitude of measured forecasting elements within the forecast lead time. The permissible forecast error of the occurrence time shall be in accordance with Table 8.4.2.

Table 8.4.2 Permissible forecast error of occurrence time of ice forecasting elements

Forecast lead time (d)	≤ 2	3 - 5	6 - 10	11 - 13	14 - 15	≥ 16
Permissible forecast error (d)	1	2	3	4	5	7

8.4.3 Ice forecasting scheme rating shall take the forecast qualified rate as the rating indicator and the rating criteria for ice forecasting schemes shall be in accordance with Table 8.4.3. The class of a forecasting scheme shall be determined based on the comprehensive consideration of various forecasting elements.

Table 8.4.3 Rating criteria for ice forecasting scheme

Class	A	B	C
Forecast qualified rate QR	$QR \geq 80.0\%$	$80.0\% > QR \geq 70.0\%$	$70.0\% > QR \geq 60.0\%$

8.5 Rating Criteria for Tide Level Forecasting Schemes

8.5.1 Tide level forecasting scheme rating shall assess the forecasting effect of the forecasting elements and their occurrence time, including the normal tidal level, maximum water surge, and maximum tidal level.

8.5.2 The permissible forecast errors of tide level forecasts shall be taken according to the following requirements:

1. The permissible forecast error of the normal tide level (high tide level and low tide level) takes 0.30 m.

2. The permissible forecast error of the maximum water surge of storm tide takes 20 % of water surge, with 0.75 m as the upper limit and 0.10 m as the lower limit.

3. The permissible forecast error of maximum tide level of storm surge shall be calculated by the formula:

$$\delta = 0.0577 H \sqrt{\Delta t} + 0.15 \tag{8.5.2}$$

   where

   δ is the permissible forecast error (m), taking 1.00 m as the upper limit and 0.15 m as the lower limit;

   H is the water surge at the measured maximum tide level (m);

   Δt is the forecast lead time (h).

4. The permissible forecast error of occurrence time of tide level shall be taken according to the tide type: for semi-diurnal tides and mixed tides, it takes 0.5 h for normal tides and 1.0 h for the maximum water surge and high tide of storm surge; for diurnal tides, it takes 1.0 h for normal tides, 2.0 h for the maximum water surge, and 1.5 h for the high tide of storm surge.

8.5.3 Forecasting scheme rating shall use the forecast qualified rate as the indicator. The rating criteria for tide level forecasting schemes shall be in accordance with Table 8.5.3 and the class of schemes shall be determined based on the comprehensive consideration of various forecasting elements.

Table 8.5.3 Rating criteria for tide level forecasting schemes

Class	A	B	C
Forecast qualified rate QR	$QR \geq 80.0\%$	$80.0\% > QR \geq 70.0\%$	$70.0\% > QR \geq 60.0\%$

9 Operational Forecasting

9.1 Hydrological Station Network and Water Regime Data

9.1.1 The hydrological data used in operational forecasting shall meet the requirements of the operational forecasting scheme. The hydrological data acquisition and transmission shall be automated for the hydrological stations. When automatic telemetry is not possible, manual measurement and reporting may be adopted.

9.1.2 The hydrological station network shall be maintained stable. Timely adjustment shall be made in one of the following cases:

1   Significant change occurs to the water regime due to the change in natural conditions or the effects of human activities.

2   The hydrological stations need to be adjusted if required by the change in hydrological forecasting scheme.

3   The hydrological stations fail to fulfill their intended function due to the change in measurement conditions.

9.1.3 The water regime data reporting interval shall meet the requirements for operational forecasting. For a manual station, the reporting requirements shall be determined according to the forecasting scheme.

9.1.4 Effective measures shall be taken to guarantee the accuracy and timeliness of the data used for operational forecasting. The hydrological data acquisition and transmission shall meet the following requirements:

1   The hydrological data acquisition shall meet the technical requirements for hydrometry.

2   The automatic data acquisition and transmission shall comply with the current sector standard NB/T 35003, *Technical Specification for Hydrologic Telemetry System of Hydropower Projects*.

3   The manual data measurement and reporting shall be undertaken by a designated person. The data reporting shall be carried out according to the reporting requirements.

9.2 Requirements for Organization and Staffing

9.2.1 A water regime division shall be established during the project construction and operation periods, responsible for the hydrological measurement and forecasting.

9.2.2 Technical staff for hydrological work of the project shall be reasonably assigned to ensure the 24/7 hydrological operation during flood season and uninterrupted hydrological operation around the year.

9.2.3 Technical staff for hydrological operation shall:

1 Be familiar with hydrological measurement and forecasting, and experienced.

2 Be familiar with the physical, geographical and hydrological characteristics of the river basin in the region, as well as the construction and management of hydropower projects.

3 Understand the hydrological station network and the hydrological forecasting scheme for the river basin.

4 Understand the historical floods, droughts and water logging disasters and hydro-meteorological evolution pattern in the region.

9.3 Software Systems and Forecasting Operations

9.3.1 Operational forecasting shall use software systems and, if necessary, resort to related tables and graphs of rainfall-runoff correlation, discharge correlation, stage-discharge correlation and rainstorm isogram. The operational forecasting software systems shall have the following functions:

1 Realize connection to the real time database, data retrieval and data processing.

2 Realize the process of the forecasting scheme and carry out forecasting analysis and calculation.

3 Display and output the forecasting results.

4 Carry out real time correction and interaction.

5 Assess the accuracy of operational forecasts.

6 Manage historical hydrological data and forecasting results.

7 Adjust and optimize model parameters.

8 Allow for conferencing and decision-making on the forecasting results.

9.3.2 The software systems for operational forecasting shall comply with the relevant national and sector standards concerning data processing and computer software development.

9.3.3 Before the software systems for operational forecasting are put into operation, they shall be tested and verified against the requirements of the corresponding forecasting scheme.

9.3.4 The operational forecasting shall adopt a number of methods. The forecasting results shall be analyzed comprehensively before the final results are proposed.

9.4 Information Reporting and Dissemination

9.4.1 The competent water regime authority for the project shall establish a system for water regime information reporting and dissemination.

9.4.2 The water regime division shall report the water regime information to the competent water regime authority. An accountability system shall be implemented for water regime information reporting or dissemination.

9.4.3 Water regime information shall be reported and disseminated at regular interval on a rolling basis according to the requirements for water regime information during project construction or operation.

9.4.4 When the project safety is severely threatened by a flood and a sudden change occurs to the water regime, which possibly have impact on the flood control for the area beyond the project, the water regime division shall request the competent water regime authority to report to the local water authority and the government.

9.4.5 The water regime information may be reported and disseminated in the form of paper documents, fax, telephone, SMS or network, and the reporting of important water regime information must be acknowledged by the recipient.

9.4.6 In the case of any significant change in climate and water regime or considerable difference between the forecast water regime and the measured water regime, the corrected forecast shall be disseminated in time.

9.5 Forecast Accuracy Evaluation

9.5.1 Forecast accuracy evaluation shall cover the accuracy of forecasting elements of individual operational forecast, the overall forecasting effect and, the timeliness of flood peak forecasting.

9.5.2 The accuracy of forecasting elements of individual operational forecast shall be rated as excellent, good, pass, or fail. The evaluation criteria for the accuracy of forecasting elements shall be in accordance with Table 9.5.2.

Table 9.5.2　Evaluation criteria for accuracy of forecasting elements

Accuracy class	Excellent	Good	Pass	Fail
Ratio of forecast error to permissible error E	$E \leq 25.0\%$	$25.0\% < E \leq 50.0\%$	$50.0\% < E \leq 100.0\%$	$E > 100.0\%$

9.5.3 The overall evaluation on the forecasting effect shall use excellent rate, good rate and pass rate as the indicators. The indicators are subjected to the statistical calculation on the percentage of the number of forecasts with an accuracy better than or equal to the accuracy level of forecasting elements in the total number of forecasts within a given period of time.

9.5.4 The timeliness of peak flood forecasts shall be classified into levels A, B and C. The timeliness level evaluation of peak flood forecasts shall be in accordance with Table 9.5.4. The coefficient of timeliness shall be calculated by the following formula:

$$CET = \frac{EPF}{TEP} \qquad (9.5.4)$$

where

CET is the coefficient of timeliness, rounded to two decimal places;

EPF is the effective lead time (h);

TEP is the theoretical lead time (h).

Table 9.5.4 Timeliness level evaluation of flood peak forecasts

Timeliness level	A	B	C
Coefficient of timeliness CET	$CET \geq 0.95$	$0.95 > CET \geq 0.85$	$0.85 > CET \geq 0.70$

9.6 Forecast Analysis and Summary

9.6.1 The water regime division shall submit the annual or staged analysis and summary report to the competent water regime authority. The report should contain:

1 Rainfall and water regime.

2 Acquisition of water regime data.

3 Operational forecasting and information dissemination.

4 Evaluation on the accuracy of annual or staged forecasting.

5 Good flood forecasting practice.

6 Identification of the cause of forecast error.

7 Suggestions for modifying the forecasting scheme if necessary.

Appendix A Contents of a Flood Forecasting Scheme

1 General

1.1 River Basin Overview

1.2 Hydrometeorological Characteristics

1.3 Project Overview and Forecasting Demand Analysis

2 Hydrometeorological Data

2.1 Hydrological and Meteorological Station Network

2.2 Data Processing and Analysis

2.3 Flood Travel Time Analysis

3 Flood Forecasting Model

3.1 Model Selection

3.2 Model Structure and Principle

3.3 Model Parameters

4 Flood Forecasting Scheme

4.1 Scheme Formulation

4.2 Model Parameter Calibration

4.3 Scheme Rating

4.4 Scheme Verification

4.5 Result Analysis

5 Instructions to the Use of Flood Forecasting Scheme

Attached Tables

Attached Drawings

Appendix B Contents of a Runoff Forecasting Scheme

1 General

1.1 River Basin Overview

1.2 Hydrometeorological Characteristics

1.3 Project Overview and Forecasting Demand Analysis

2 Hydrometeorological Data

2.1 Hydrological and Meteorological Station Network

2.2 Data Processing and Analysis

3 Runoff Forecasting Model

3.1 Model Selection

3.2 Model Structure and Principle

3.3 Model Parameters

4 Runoff Forecasting Scheme

4.1 Scheme Formulation

4.2 Model Parameter Calibration

4.3 Scheme Rating

4.4 Scheme Verification

4.5 Result Analysis

5 Instructions to the Use of Runoff Forecasting Scheme

Attached Tables

Attached Drawings

Explanation of Wording in This Code

1. Words used for different degrees of strictness are explained as follows in order to mark the differences in executing the requirements in this code.

    1) Words denoting a very strict or mandatory requirement:

        "Must" is used for affirmation; "must not" for negation.

    2) Words denoting a strict requirement under normal conditions:

        "Shall" is used for affirmation; "shall not" for negation.

    3) Words denoting a permission of a slight choice or an indication of the most suitable choice when conditions permit:

        "Should" is used for affirmation; "should not" for negation.

    4) "May" is used to express the option available, sometimes with the conditional permit.

2. "Shall meet the requirements of..." or "shall comply with..." is used in this code to indicate that it is necessary to comply with the requirements stipulated in other relative standards and codes.

List of Quoted Standards

NB/T 35003, *Technical Specification for Hydrologic Telemetry System of Hydropower Projects*